Colour in Industrial Design

Dale Russell

The Design Council

Published in the United Kingdom by
The Design Council
28 Haymarket
London SW1Y 4SU

Typeset in the United Kingdom by TDR Photoset Ltd
Printed and bound in the United Kingdom by Bourne Press Ltd, Bournemouth
Designed by Nicole Griffin

British Library Cataloguing in Publication Data

Russell, Dale
Colour in industrial design. – (Issues in design).
1. Industrial design. Use of colour
I. Title II. Design Council III. Series
745.2

ISBN 0 85072 283 7

Picture credits: The Design Museum: p.9 (both); Ian McKell (photographer), Reay Keating Hamer (agency): p12; Braun UK: p.13; Addis Ltd: p.33 (both); Kenneth Grange, Pentagram: p.36; Kodak Ltd 1990 TV advertising for KODAK 'Kodacolour' GOLD Film: p.37; Victoria & Albert Museum: p.40 (top); Sensiq: p.57; Pattie Boyd: p.60; Newell and Sorrell Ltd: p.64 (both); Giant Ltd: pp.84-85; Hydro Polymers Ltd: p.88.

Contents

Preface

In writing *Colour in Industrial Design* I have attempted not only to inspire good colour use and make designers more aware of the reasons behind colour selection but also to inform management about the role of colour specification in design. My approach has been to comment on relevant issues and analyse case studies rather than teach colour application. However, this book is only the tip of the iceberg, and its aim is to provoke thought and stimulate interest.

I firmly believe that colour should be taught across the disciplines, allowing for a free flow of information regarding materials, technology and inspiration and I am grateful to Central St Martin's College of Art and Design and the Royal College of Art, amongst others, for allowing me to teach colour in context alongside theory. I would like to thank Roger Sale, Steven Kyffin, John Wood and Hugh Gilbert at the RCA. I am grateful to the following, without whom this book would not exist: Teresa Collins, Angela Gillic, Kenneth Grange, Alan Herron, Richard Jenkins, Frances Newell and Ken Rees.

Finally, I wish to thank my husband Steve and my daughter Luce Scarlett for their patience, support and encouragement.

Dale Russell

December 1990

1 Introduction

Colour specification

Colour is the Cinderella of the industrial design process. Often treated as an afterthought, colour specification is rarely even mentioned at the briefing stage. It is also extremely unusual for a designer to retain the original colour specifications relating integrally with the form and technology of the product, throughout the many processes of manufacture. Variations on 'We had a specialist team of designers working on the project but the chairman's wife chose the colour' are frequently heard (whether in exasperation or as a witty anecdote), revealing the prevalent archaic and insensitive attitude to colour within industrial design. There are, of course, exceptions to the rule, in the form of excellent existing designs (such as the Stanley screwdriver) that have incorporated colour, form and function throughout the process of design and manufacture. However, there is a long way to go before colour can be considered an integral part of design alongside form, function, technology and finance.

It is necessary when looking at colour in industrial design, to observe the wide range of technologies and attitudes across the various disciplines, while avoiding sweeping generalizations. This book attempts to examine the roles of those responsible for colour specification within industrial design and its related areas.

An awareness of colour specification and its relating logistics across the disciplines can provide an insight into more specific fields of industrial design as well as broadening the concept as a whole. As John Heskett says:

> It [industrial design] can frequently overlap with other areas of design, indeed, practitioners have claimed its range of concern extends from 'a lipstick to a steamship'. (Conway 1987)

Colour in Industrial Design encompasses 'grey' areas which include:

- Furniture: for its relationship with contract and domestic products.
- Graphics: whose final colour specification forms a corporate and public identity – revealing the image of the product through packaging, exhibitions, annual reports and all forms of literature.
- Corporate identity: relating to product, graphic, interior and exterior design, as well as to peripheral areas such as the design of staff uniforms. 'Everything that the organisation does must be an affirmation of identity' (Olins 1989).
- Cosmetics: whose design and manufacturing processes involve a diverse range of methods and materials relative to those of industrial design.

The various processes of colour specification in industrial design are performed not only by the designer, but also the design engineer, colour specifier, product manager or marketeer. For the purposes of this book, however, the term 'designer' will generally be employed, as it is the role of the designer, and their relationship with colour within industrial design, that is being evaluated.

The book observes and analyses the various processes performed by the designer, whether they work as a product designer or in associated areas. The designer retains a key role in making products that have identity and are aesthetically pleasing, through the function of colour. Although

Above: Ericsson DB 1001 telephone by Jean Heidberg (1931): through colour, a functional product becomes part of a decorative interior

Left: The reflective chrome surface enhances the form of a Russell Hobbs kettle

designers are not automatically involved at the conception of a product, and few make the final colour decision, they should all be aware of the creative, manipulative and financial power of colour.

The freelancer often plays a major role in colour specification. Many companies use freelance colour consultants and designers to create colour ranges for existing products, to introduce new products into existing ranges, to give new vigour to existing ranges, to colour new ranges and, occasionally, to design and colour new products. Larger organizations often keep a freelancer on a retainer, so that they are able to build a relationship that is beneficial to both the company and the freelancer. It is more common, however, for the freelancer to be employed either to supplement the in-house design team, or to work with industrial engineers and technologists in companies without designers.

The freelance colour specifier specializes in the colour, texture, surface and finish of a product. They work on the appearance of all kinds of products from textiles to hard plastics, whether using natural finishes with the addition of a colour tint, or creating completely new finishes, including the examination of pigments and resins. Freelancers may also work on existing ranges, and the most interesting projects can often be found within these limitations.

Forming the chain

Different products work on differing lead times. Although the average design process for a product takes 18 months to two years, a designer dealing with small products might work on a relatively short, ten-month turnround, while the car industry demands a lead time of approximately four years. Consequently, the automotive designer is working on a lead time that is further into the future than fashion, which is normally seen to be the 'crystal ball' in colour prediction. Like cosmetics, the car is thought to be an accessory – an extension of the purchaser's identity – and so a 'chain' is formed. Although the products are basically unrelated, there are significant points of contact for colour specifiers.

A speaker from Merck at a Colour Group (see page 96) seminar (December 1986) discussed new technologies relating to pearlescents, with a multi-discipline audience of designers. Following the seminar, four designers from contrasting disciplines (automotives, plastics, hard flooring and cosmetics) proceeded to interpret pearlescents into their respective media.

When working in industrial design, a designer must overcome the problem of specifying colour across different media. A particular shade of red used with plastic will never look the same when applied to fabric. Even if the colour reads as an identical shade, when applied as a smooth gloss on metal or as the tufting of a carpet, it completely changes character. Although spectrophotometers, tristimulus colourimeters and sophisticated computer systems are used for colour measurement, there is still a place within the system for the trained human eye. The calculated readings will take into account the related texture, giving accuracy, objectivity and quantitative results repeatable over long periods of time, but the human component allows for perception of the warmth and coldness of a shade. Therefore, the most accurate method of colour specifying across different media combines the emotive visualization with a true colour image.

When working with colour, the designer must have the ability to communicate specific colour reference to others. There are many systems for specific referencing, such as Pantone (ink on paper swatches, for printing, and fabric swatches for fashion) and Chromatone (tufted yarn, used in the carpet industry). A palette relating to specific areas of industrial design is catered for by systems that include British Standards Institution and RAL German Institute for Quality Assurance and Labelling, Reg. A. The Munsell Color System, Colourcurve from the USA and the Natural Colour System – developed in Sweden at the Institute of Stockholm, and under licence in the UK as Colour Dimensions in collaboration with Dulux and ICI – are all notation systems. In theory these systems cross the disciplines with their ability to notate a colour, regardless of material. These systems, plus many more, are crucial to the specifier dealing with the elaborate process of industrial design. 'Unlike design for ceramics, glass or textiles, it [industrial design] is not confined to one market' (Heskett 1980). The flat effect of ink on paper, the milky-white bias of a paint system, or the restricted palette of a specific notation system can become destructive when used in the wrong context, providing a distorted or restrictive palette for the designer. Therefore, access to and a knowledge of various colour specifying systems is of enormous benefit to the designer.

Red Stripe lager advertisements use a red stripe to create a literal and atmospheric product image

It's useless to create a design in glowing colour if the process by which it will be reproduced cannot give similar results.
(Jones 1950)

Altered images

It is significantly cheaper to change the colour of a product than to redesign it. The look and feel of a product can be altered by using different forms of surface effect, such as matt, gloss, hammerite or mica. The proportional use of colour is also a vital constituent in colour specification as the dimensions of a product can be redefined if the proportional use of colour is changed. The ability to change the visual perception of a particular angle or curve can be brought to dramatic – or subtle – effect with the highlighting or mass coverage of body colour. The use of complementary colours can confirm the strength of chroma, while strategically placed contrasting shades can define, or redefine, shape. The character of the design's form or function can be enhanced by a shade which reflects its performance, components or use. For example, the use of burgundy for a vacuum cleaner gives the product a pleasing appearance, suitable to be

Braun used black on the Flex Control shaver to enhance the form of the product

left in a living area rather than hidden in a broom cupboard; or the use of black in product design, so that light reflection complements the form. The sophisticated use of black in this way is evident in Braun products.

It is also imperative to consider lighting while specifying colour, as the effects of lighting can distort or project a colour with equal power. The distortion of colour between different forms of lighting (known as metamerism) must always be considered, as a product must retain its true colour in artificial light at the point of sale, as well as in natural light when bought by the consumer.

Colour, combined with light and surface effect, can effectively control shape and form. It is important that colour is not regarded simply as a piece of flat pigment forming a hue, but is related to the type of material and application. The actual shade is sometimes subordinate to the tonal effect, which can give a range of products a harmonious 'family' of colours, or endows a product with a subtle and classic sophistication. As new technology provides more ways of applying colour, whether waterborne, solvent held or with wet or dry application, technology also demands different colour functions, such as light/colour-reactive solar energy and light/colour-reactive camouflage.

Psychological reactions

The psychological, physiological and cultural effects of colour should also be taken into account when specifying colour for a product. Colour can 'control' heat: white reflects colour while black absorbs it; and red psychologically creates the feeling of warmth, as well as physiologically creating a reaction that leads to a faster rate of human blinking. The primitive instinct associates red with fire and blood:

> Respiratory movement increased during exposure to red light and decreased during blue illumination ... frequency of eye-blinks increased with exposure of red light and decreased with blue light. (Gerard 1957)

Theoretically, chrome-yellow is an obvious choice as a base area, as yellow is the most visible colour and provides a striking contrast to black, thereby allowing the greatest possible degree of visibility. However:

> ... nature has created a danger warning system based on the combination of yellow and black. Some species of bee, wasp, frog and snake have yellow and black stripes to warn that they are poisonous. In many parts of the world, humans have adopted this warning signal ... for areas where there is poison, toxic fumes or waste, or radiation. (Russell 1990)

Therefore, the combination, pattern and proportion of these two colours should be considered carefully.

Global design

Modern, sophisticated methods of communication allow access to almost anywhere in the world, and only lack of funds in particular areas prevents the completion of an instant global communication network of satellite, telephone, fax and computer. Supersonic travel has enabled the Atlantic to assume the proportions of the English Channel, allowing someone to attend meetings in London, New York and Paris in one day. Although there are obviously vast cultural, political and financial differences between countries and continents across the world, this 'instant' global accessibility has created a form of inter-global design where products (such

My First Sony: produced in primary colours to educate and appeal to children

as the Sony Walkman) can be sold worldwide without design or colour change. It has also created a form of cross-fertilization, with Italian design having an effect on the British, whether this is in the form of a continuation of Memphis colours or the interpretations of the Domus Academy; or the Japanese showing that it was not only Henry Ford who felt 'any colour is fine as long as it is black' as they manufacture their mass-produced wares to fit a global culture.

Alongside a global or pan-European form of design is a sense of design heritage that creates a defined geographical and cultural barrier. Greater emphasis is increasingly placed upon the regional and historical aspects of design, in order to contrast with the mass manufacture of global products. The return of the artisan designer and craftsperson, along with a growing appreciation of ethnic products, are signs that traditional design is still relevant in contemporary society.

Differing cultures traditionally assign diverse meanings and associations to colours: white represents innocent purity and is worn by brides in western countries, while in China it is the traditional colour of mourning; black signifies death in the west and is worn for mourning, while in Egypt it represents re-birth. Another factor to be considered is regional 'personality' and reaction to colour. It has been observed that a red or white car is generally suited to the fiery temperament associated with Mediterranean peoples, while the stoic, less passionate Germans would usually prefer a black car.

If a product is to be sold globally it is therefore highly relevant to take into account the cultural background of each country.

This leads to the area of personal selection of colour; no amount of theorizing on colour choice will prevent the individual from selecting a product in a specific colour 'just because they like it'. The reason why racing green rather than canary yellow should be chosen for a sports car can be explained in this way.

Market research shows that while only a small percentage selected green from a random screening of colours, a large proportion of prospective buyers chose green. The real reasoning behind the colour selection is that the designer felt that the particular shade of green related to the shape and character of the particular model of car. The same reason would apply to the selection by prospective purchasers, as the image created by outward appearance is almost as important as the actual function.

Market research

A key factor in colour choice is market research, and although this is referred to in this book in individual case studies, it is still important to include a separate analysis on a market research company which specializes in colour (see Chapter 7). Market research is used to 'sell' the product – and this includes the colour – to clients and manufacturers, marketing teams, sales teams and store buyers as well as the consumer.

Market research and marketing are still often regarded as enemies by the designer when dealing with creative colour choice. Marketing is reactive, not proactive; the information is gleaned from a buying public that is basing its reaction on an existing colour palette rather than unknown colours of the future. A market research department receives information from management when formulating questionnaires, and very few people in management understand colour. The designer needs to be fully briefed by marketing colleagues and allowed to help formulate questions for market research information so that a presentation is given without forming a battleground on which the innovative colours are likely to fall.

If colour specification was included at the conception of a product and given support from management and marketing, it would allow greater calculated freedom, so that a 'safe' palette could be created. This would allow the addition of a few innovative colours that are calculated loss leaders, or the creation of a basic range of core colours that tonally complement one another but allow the inclusion of new and complementary random colours. If a designer chooses a colour palette which is considered too advanced and innovative, a palette of relating 'safe' colours can be brought in to prepare for the arrival of these 'future' colours; hence, the need for up-to-date background market research materials.

It is necessary for the designer to look at the way the public approaches design and then relate to the general feeling of the consumer, rather than just relate to specific colours. For example, if the economy is slow, the colours coming in will be tonal and more neutral, reflecting the mood of the market. The colour choice will then relate to the feeling that the age of fast consumerism is receding and that a more classic look is becoming popular. This, in turn, will give products a 'longer-lasting' appearance.

People are becoming used to choosing colour and are more sophisticated in their approach. They are aware that there is more than one shade of mid-grey and are combining colours in new and more interesting ways. The approach to colour has completely changed, and although the colour palette is returning to soft browns, for instance, it is not the same palette of browns as it was the last time around. There is much more colour available, so the role of the colourist is even more important.

'Safety' is a watch word in industrial design, and market research runs in tandem with the cause of sensible, statistically sound colour ranges. There is, however, a case to be made for the outsider or loss leader in colour terms. Creative use of colour can lead to public acknowledgement of good design, to media coverage and to design awards. It can draw attention to a new (or stale) product or to a new category of consumer, provide a sales theme or enhance a brand image.

There is a remote chance that the outsider actually becomes a steady seller and builds a reputation over several years. Situations also emerge in which the colours have been ahead of their time and a time gap has to be created to allow public taste and/or fashion to catch up.

Colour theory

Although the practicalities of colour specification have been observed, it is important that the essence of colour itself is not lost. An advantage of colour is that its possibilities are infinite. A comparison to music has been made by many, including Kandinski: '... representing the different blues in terms of music ...' (Kandinski 1912), Newton, Field, Goethe and Wagner, with compositions notated by Castel. It is therefore quite sensible to talk of colour notation. However, when dealing with manufacturing processes it is necessary to consider the omnipresent factor of finance. It is not practical to compose an avant-garde opus in colour when specifying a multimillion-pound range of small domestic products.

Art-school-educated designers will have studied colour theory. Itten's theory (Itten 1970), taught with the aid of coloured squares, emphasizes colour awareness, the qualities of colour and the varieties of possible effect. This process of teaching an appreciation of colour temperature and optics through warm, cold and complementary colours gives a basic grounding that is invaluable to a designer. There is, however, a need for lateral teaching that builds on this theory by incorporating function, colour, aesthetics and context within the concept of a design.

Nevertheless, students who receive the most basic art school tuition are advantaged when compared to the plight of a designer trained in industrial engineering without any colour training, or the head of department who has progressed from apprenticeship through the company, having always seen colour as an afterthought. The truth is that many designers and engineers are specifying product colour without any formal knowledge of colour. The roles are reversed when a colour technologist has the greatest knowledge of colour theory and science but has little desire to work with the aesthetics of tone and hue. It is quite possible to be colour blind and still work with colour as a chemist, because colour at that level becomes a formula rather than an integral part of form and function.

The ability to assemble colour in an appropriate way is instinctive, and it is my firm conviction that colour teaching cannot make a good colour specifier without this instinctive 'gut' reaction. It is frightening, however, to discover how many people involved in colour specification throughout the manufacturing process of a product admit that they have no – or very little – colour sense.

The power of colour

While industry is gradually becoming aware that colour should play a more significant role in the processes of design and manufacture, the public is also becoming more demanding and discriminate in its views on the way colour is handled in product design. The media, and advertising in particular, have brought colour to public attention: this public awareness creates an atmosphere that is felt throughout the design industry and is reflected in the way consumers select their purchases.

Not everyone is naturally colour literate, but most people have become more aware of style, and related colour use, through the abundance of style and design magazines and television. A magazine such as *The World of Interiors* was once a minority interest magazine purchased from a few shops specializing in design, but this, and other specialist design publications, can now be found in most newsagents.

Colour trends are greatly influenced by advertisements, whether on television, film, hoardings or in magazines. Advertisements are able to exploit colour in various ways: the Lynx anti-fur campaign (1987) promoted a known colour association – red signifying blood; Red Stripe lager (August 1987) adopted a minimalist approach to colour by using black and white with just a touch of colour to create a strong and evocative impact; and Sunkist (March 1989) focused on a specific colour (orange) to create the imagery for the product (an orange-flavoured fizzy drink).

The power of colour imagery, however, is best illustrated in the advertising campaigns for cigarettes. Here, the copy (apart from the health warning) is replaced by the simple, but effective, use of colour.

The innovator of these campaigns was Benson & Hedges, with its use of gold (the colour of the packet) in the various 'disguised packet' advertisements. Then Silk Cut unfolded the sophisticated campaign in which the public were 'conditioned', through a series of advertisements, to respond to the colour purple and a simple reference to a cutting device. The campaign was launched using the image of a piece of purple silk, either with a loaf of cut bread and a knife, or wound around scissors. The campaign also included advertisements consisting of a purple-silk-upholstered barber's chair with a poodle and hair clippings; followed by the purple-seated chair with just the clippings. Perhaps the most subtle and sophisticated of the Silk Cut advertisements, however, was the image of a purple shower-curtain with the silhouette of a shower-head. The observer had to translate this image into a reference to the film *Psycho*; a feeling of having passed a MENSA examination allows the image to be retained with pleasure. The only clue was purple! This form of advertising is slowly attracting product manufacturers, while there is a growing use of product imagery through colour that allows the actual product to stand forward, for example Kodak film and Berol pens.

The media have not only brought colour to the attention of the public, but also the manufacturer. Paint companies now not only advertise the actual paint but instruct how to apply a paint finish. The advent of paint mixing, from companies such as Dulux Colour Dimensions and Crown Colour Expressions, means that individual selection is available not only through interior design specialists serving an exclusive market, but to all levels of society at the local DIY store. The freedom to obtain a specific shade has allowed a subtlety of colour decision that, in turn, leads to a demand for greater sophistication of product colour choice.

Computers are also being employed in conjunction with paint mixing and supply, allowing the consumer to 'paint' an image that represents their home or workplace. This device gives the consumer the courage to try new colour combinations as they feel the mistakes are being made 'on screen'. It can also show how to highlight different areas within a room, such as mantlepiece, dado and cornice. These computer programs often contain suggestions developed by interior designers, so the more cautious can follow advice. This does not necessarily lead to better colour use, but it does make the consumer feel as though they deserve greater freedom of colour choice within other areas such as product design.

In the USA 'a large toy manufacturer has 6-12-year-old children working in tandem with a computer that manufactures toys' (Ogden 1990). The children design and colour the toys alongside the manufacturing process, and the prototype is completed in a few hours. The project has, so far, produced successful innovative designs incorporating colour.

Colour is being acknowledged for its power to create a mood or image. Bottle green, mulberry and navy, for example, can all evoke a bespoke quality appearance. Black and white evoke the image of 'modern' design – whether in the 1920s or the 1980s. There has been much publicity surrounding the way that colour has vibrantly emerged from a chrysalis in the 1990s; the hard and materialistic image of the 1980s, exemplified by monochromatic shades, is being replaced by a soft and caring attitude. Alongside the emergence of a full colour palette, the colours at the forefront of this new era are taken from natural elements like sand, granite, marble, lichen and wood, yet white is the colour chosen to represent the 'New Age'. This, of course, is hypocritical, as the bleaching processes, along with the extra energy used in laundering, are ecologically unsound.

Ecology

Ecological matters and the impending Single European Market are closely-connected issues. Much legislation currently being adopted by the UK to enable it to become an active member of the EC in a united Europe is closely connected to environmental issues. Legislation involving colour specification includes the DIN (Deutsches Institut für Normung eV) standard and the banning of certain colour pigments and additives in food. The DIN standard relates to colour, reflectance, saturation and reflection, and defines light values, putting forward a 20–50 per cent limit on light reflectance. It was enforced as a legal requirement in Germany after insurance companies found that thoughtless colour specification alongside poor lighting in offices caused sickness amongst workers.

The arrival of the Single European Market will herald changes in UK rules regarding food products: the legislation banning harmful colour additives in food will help to change our perception of colour. Expectations of food colouring will change as we are surrounded by 'natural' coloured food as opposed to the highly-dyed green pea and pink sausage we, in the UK at least, have grown up with in the past. Packaging too will change as the public is taught to find these 'new' dull-coloured foods appetizing; packaging in supermarkets will become less brash, with tertiary colours being accepted into what was a very narrow palette of primary colours. Babies' disposable nappies no longer have to appear whiter than white: colourful shades relating to babies' skintones or clothing have been introduced, giving aesthetic choice while no longer necessitating the harmful bleaching process that was threatening to the child's future.

Although our attitudes to basic colour representation are starting to change, many designers seem to believe that ecologically sound pigments and dyes for 'green' packaging and products have to appear in dull, muted shades. It is almost felt that a 'hairshirt' of sombre colours should make consumers feel environmentally sound, while having fun with primary and secondary colours is somehow decadent. Nature provided the perfect colour system with the banana. The outer casing forms a colour code, from green to black, revealing the state of the fruit inside; the most optimistic and friendly shade – yellow – announces that the product is ready to be eaten; the gentle shade of the inner, protective packaging relates to the fruit itself; and the entire object is biodegradable.

The evolutionary process of colour usage can be compared to a video: it is on fastforward and rewind, with the pause-button about to be pushed. As daily revelations on environmentally sound processes appear, confusion increases. While it is generally agreed that a holistic approach must be taken, debate around the issues of recycling versus technological innovation in areas such as colour application, chemical potency and the use of energy continually rages. Do we fastforward into a recycled, de-inked 'green' approach, or should we rewind our memories to the days of carefully saved brown paper and rewound balls of string, and pulped cardboard egg boxes easily stacked waiting for re-use? There is no reason why 'safe' colour usage could not be applied in these basic terms.

Footnote

It remains to be said that colour is still considered to be the outcast of industrial design; however, a greater social awareness of colour, through technological, economic, political and environmental issues, is making manufacturers and management more aware of the importance of placing colour alongside form and function in industrial design.

2 Addis

Housewares

The kitchen is an area in which modern and functional design has always been accepted with enthusiasm.

> Homes in the 1930s were filled with brightly coloured, compression moulded eggcups ... attractive light-coloured domestic plastic had at last arrived. (Katz 1984)

The mass appearance of plastic housewares was immediately seen as a tremendous innovation, as the products were hygenic, cheap and colourful.

Richard Jenkins is the design director for Addis Limited. He joined the company in 1968, and has since been involved in the design and development of the majority of its products. He broadened his scope of activity to cover all aspects of the company's design output (for example, packaging).

Trained in industrial design, Richard enjoys the challenge of creating attractive, well-made, functional and cost-effective products. His designs are influenced more by the search for the simplest solution and the classic line, than by transitional fashions. Richard finds that when working in the area of mass-produced consumer goods it is often difficult to find the balance between new and exciting, yet tasteful products.

Aesthetically, colour rather than form is probably the prime consideration for consumers of Addis products. High-quality manufacture, and good design and function are of course essential, but according to Richard: 'colour is the key to the emotions and taste of the average shopper'.

Since the range of products marketed by Addis encompasses not only housewares but dental care, beauty products and others, the colour decisions are made in the light of many consumer and research influences.

Brand awareness

Market research (Addis 1988) revealed the following statistics with regard to consumer recognition of brand names within the housewares market: Addis – 97%; Plysu – 31%; Curver – 13%; Stewart – 12%.

Richard explains how one product evolved:

> In 1957 Addis introduced the plastic kitchen pedal bin, a small round bin designed to take the modest amount of kitchen rubbish which the average family threw out. Over the past 30 years, the size and design of Addis kitchen bins has changed significantly. Today, Addis's best selling kitchen bin is over four times the size of the bin of the 1950s, and is used to dispose of a plethora of plastic, polythene, polystyrene ...

Addis was also one of the earliest companies to produce colour blocking: in the early 1960s they had ranges consisting of reds, mid- to light-green and mid- to light-blue. Taking this colour-led approach, Addis manufactured a range of products related only by colour.

Product colour

Colour inspiration at Addis tends to come from an awareness of colour trends and general issues throughout Europe and other parts of the world.

Noting the evolution of colours over a period of time is a firm foundation for projecting new colour trends. Richard believes that the most difficult trends to lead are the rapid, deliberate shifts from one palette to another: 'This is where pure inspiration comes in and here there is no real substitute for instinct'. In certain areas of design, such as fashion, it is relatively easy to target a small sector of the market (for example, teenagers) and make an accurate colour selection. In the mass-produced housewares market, however, it is more difficult to determine such defined sections and so specify appropriate colour trends.

After many years' experience, Richard realizes that it is important to have many more ideas than you can hope to use and to discover to what extent the consumer and the marketplace can be led. Working too far ahead will alienate the consumer, so the timing of colours is crucial.

At present, the kitchenware colour palette is subdued and limited as a result of the dominance of wood in kitchen units. The preference for colour toning rather than contrasting has led to a revival of warm neutrals. Richard observes that:

> This is of course a fashion, and will in time change again. It has probably very little to do with psychology, since these movements are too widespread to be a reflection of the individual's mental state. These movements tend to occur in waves of reaction against the dominant theme of the recent past, so pastel follows primary, neutral follows pastel and next ... well that is what we are all trying to find out.

Addis have recently created a new range of housewares that still relates to these wood-inspired neutrals and the family of colours ranges from coffee and biscuit to ivory, with terracotta as the key colour.

The challenge of creating colours which are not only new and exciting, but also correlate with consumer preferences, is made more difficult by the manufacturing limitations of plastic housewares. The number of colour variants is regulated by the practicalities of retailing, like lack of storage and display space. The answer to this dilemma lies partly in dividing the range into groups of products in order to facilitate creative colour use. Some products are strongly influenced by the background colour of their environment, while other products can influence their environment. This is largely due to the market 'positioning' of the product range devised by the company rather than the researched preferences of the consumer.

Materials

Pigment is usually introduced in pellet form to the natural uncoloured pellets of polymer (chemical compound) just before they are melted and mixed for moulding. The colouring pellets consist of a mixture of pigment and polymer to ensure even distribution throughout the moulding process.

Highlight or accent colours are useful but many of the Addis products are manufactured in one piece, which limits their highlighting possibilities. Providing mix-and-match colours within a range is an interesting way to overcome this problem but the idea has so far proved to be unsuccessful.

Matching an individual colour and producing a sample for evaluation generally takes several weeks. Several shades of a particular colour are produced simultaneously for evaluation, especially when developing a new colour. This technique is more effective than researching only one shade at a time. It is impossible to ascertain the effect of the colour research until the colour can be seen on the actual product; the size, surface finish and form can all have a significant effect on the selected shade.

Elaborate colour effects can be achieved on injection moulded plastics but they are relatively few and seldom produce the even surface that consumers expect. Pearlescents, for example, create distinct flow-marks that can give a cracked appearance. Marble effects can give rise to similar problems, and can display unpredictable patterns in products of different shapes within the range.

It is common for three to four months to elapse from the initiation of a new colour idea to its first production date. Not all colours tested go into production; many more are produced as samples and are eliminated in consumer research.

There must be careful control over the number of variant products to be made, stocked and distributed. Due to the difficulties of allocating store space, the number of colour variations for a wide range of kitchenware must not exceed five or six. Consequently, Addis produce colour ranges which generally offer one colour from a given group, rather than a range of similar shades. This also facilitates consumer colour selection.

Market research

Addis are in a position to use their own staff to provide preference responses. With around 1,500 employees, Addis are able to obtain a reliable range of in-house responses to new designs and new colours. Addis also carry out national hall tests which consist of formal research target groups. Interviewees are brought together in one building, grouped according to socio-economic and geographic status, age and sex. Their responses to products within the target area (such as kitchens) are analysed by a professional market research team. Another market research technique used by Addis includes groups of six to ten people in a home-based situation. These are known as 'focus groups' and the interviewees discuss their own homes, favourite colours and domestic ideals.

Recent research carried out by Addis suggests that although certain products may be replaced in favour of a new colour, most consumers would only replace their kitchen plastics en bloc to match a new kitchen. Plastic kitchen products do not appear to be used in isolation to create a new scheme; redecoration at least, will be involved. According to Addis's in-house research, the top-selling colour in the current housewares range is ivory, while the least favourite is blue.

Although ivory is consistently a 'best seller' there are no shades that are impossible to sell, although their popularity is dictated by fashion. Consumers may select red because 'it is warmer' but when the colour becomes outdated the consumers may change to the new fashion colours.

Module 2000

The Addis 'Module 2000' storage system, which won the Housewares Industry Award for Kitchenware (September 1990) is a prime example of market 'positioning' in tandem with consumer preferences. Module 2000 consists of modular stacking and nesting containers in six sizes from 225 × 190 × 128mm to 450 × 380 × 225mm.

It was necessary for the new range to be differentiated from earlier products, which have a bright, toy-box colour style. The sophisticated and ingenious range of multi-purpose containers had to be given a colour range which communicated the superiority of the product.

A series of options was assembled for consumer research. These included tried and tested colours as 'controls' and new, richer, darker, more subtle colours as well as surface effects like pearlescents.

The focus groups displayed a marked preference for those colours

Left: Outdated colour range of Addis storage boxes

Below: Module 2000 storage containers

which elevated the perceived value, durability and sophistication of the range of products: anthracite grey, stone white and rich red. These are now included in the latest range of kitchen products (October 1990).

Ecological awareness

Now that cadmium-based colour pigments are alien to Addis's environmental ideals, it is more difficult to achieve bright yellow, orange and red shades. Organic pigments are also generally less opaque which causes problems when attempting to achieve a solid appearance. Addis, however, believe that 'this is a reasonable price to pay' for environmental safety.

An Addis survey (1988) revealed the following statistics regarding householders: 26% save vegetable waste; 53% would like to use a can bank; and 90% would like to do more recycling. Addis have therefore recently launched a 'Twin-Bin' containing two compartments for separating kitchen waste, which makes recycling easier and cleaner.

Colour plays an important role as it reminds the consumer of the task to be performed. The Twin-Bin is available in the same colour range as their other kitchen products, but has a dark green divider between the internal compartments, which is visible on the exterior. In this case the colour is intended to reinforce the purpose of the product.

3 Pentagram

Products

The Retractor razor was designed and coloured by Kenneth Grange of Pentagram for Wilkinson Sword. Although Pentagram is well known as a design organization and produces high-calibre design, it does not behave like a large company. Pentagram consists of a number of small pyramids, each managed by a partner, and each one independent of the other in terms of design. So, the company has the personal atmosphere of a small establishment, with the support of a large corporation.

The partners at Pentagram all have equal status within the company; Kenneth is the partner responsible for product design and development. In London, Kenneth has five partners who are graphic designers, each with different specialist skills, and another partner who is an architect and works mainly within interior design.

The development of a product becomes a personal affair for Kenneth; it is therefore pertinent to consider some of his attitudes and reasonings, even though they are not all relevant to this particular product. Kenneth feels that a designer not only has responsibilities to the client, and eventual user of a product, but also to his or her own development, and in turn the development of colleagues. Therefore, the relationship between the designer and their colleagues, which is regarded as outside the essential trio, is very important.

The razor

The product endured many changes of direction and development throughout a long design process, culminating in the fulfilment of one of the original objectives – cost cutting.

The 'retractor' part of the razor – the mechanism that ejects the blade for use and then retracts it back to a safe position – was devised in order to conceal the potentially dangerous part of the razor blade. The ingenious part of the design is that the blade unit – essentially a very critically made and accurate assembly of metal blades in a plastic support – is inserted into the handle housing by skilful organization of the geometry, the spacing and the shape. The blade can then be flipped in and out. This device saved the loose blade cover, and so was cost effective to produce, thereby fulfilling the economic ambition of the project.

The Retractor breaks the rules of razor design by using red

The colour

Kenneth believes that we see all forms in colour, and that it is not possible to think abstractly about form. If somebody says to him 'car tyre', for example, his response is conditioned: he cannot but think of it as a black object. Some of these pre-conditions are so intrinsic that the designer would instinctively know if it was risky to challenge them. Kenneth points out that shapes can be enhanced by some colours more than others: for example, he finds white cars unattractive, while in black they are a different type of product: the gloss, the highlights, the gleam and the dash all transform the personality of the product.

In any culture colour will have norms which should be adhered to: for years, custom dictated that domestic products should never be green. The colour was associated with bad luck, and was surrounded by prejudice and superstition: it was therefore commercially very risky to manufacture a green domestic product. One of the deep-seated prejudices of the shaving business was that red should never be used, because of its association with blood. However, Kenneth recalls that from the initial stages he had envisaged the razor as being red. The very form of the product was right for the colour red, or conversely, red was appropriate for the shape of the product. The colour was relevant to the ergonomics of the form, not in any measurable, mathematical sense, but in the way that the user is able to find the product easily and will (hopefully) gain pleasure from using it. This is

The toad in the Kodak Film advertisement appears to alter its shape through colour change

known as the user/product relationship. Kenneth would rather see a bright red razor on a shelf, waiting to be used, than he would a grey one. Although he obviously had personal reasons for choosing the colour red for the product, market research also revealed that red was proved to suit Wilkinson's market position.

Both the colour and the razor design have proved so popular that it is almost impossible to define particular consumer groups. If the product had appealed only to a particular group there would have been a slump in sales. There is a large age spectrum in that particular product area – men and women, between the ages of 15 and 80; no group would be large enough within such a huge spectrum to actually cover any losses were it unattractive to the other groups.

The Retractor is a good product and it is made by a reputable company: it is therefore allocated good store space. It is, of course, at this point that the colour is most valuable. There is always the risk, when choosing a bland colour, that the colour is not sensational at the point of sale; this awareness has contributed to the rise in colour variation, liveliness and personality variation in product design. In the home or office the product may not need to be sensational, but it is this eye-catching feature that attracts a consumer to buy. Using this criterion, Kenneth gave a number of good marketing reasons to back-up his colour choice. The immediate reaction to the colour by the client's chief executive was 'Over my dead body', although he was very enthusiastic when research revealed that red would be successful. One reason for his dissent was associated with the fact that Gillette dominated the market with a blue razor. Wilkinson were faced with the certainty of being number two in the market, and a dilemma: should they follow an accepted colour norm, or should they establish their own personality? Kenneth is aware that the design reaction is to establish your own identity but in the cold, commercial marketing world, there is a theory suggesting that it is financially more viable to follow the market leader. Blue was proposed to be a more appropriate colour choice because Gillette had established the relationship between the colour blue and a publicly-acknowledged, quality, reliable, value-for-money, disposable razor. Wilkinson's product, however, serves a dual purpose: it is successful in the user's hands as well as being eye-catching on the display pegs because it is attractive and novel.

Materials

The plastic used to create the Wilkinson Sword razor is a mixture of crystal and standard polystyrene. Plastic materials are originally coloured according to the chemical construction of the product; in this case, the material has a dull opal colour. Producing a desired colour for a product is achievable either by colouring the plastic at its original factory (but only if the plastic is produced in financially feasible, large quantities) or by buying the moulder in the neutral state and adding the colour at a later stage. In this case the colour was added during injection moulding. There has been a lot of development in this area, and many manufacturers have gained financially, because the material maker can produce a standard colour to be distributed worldwide, thereby lessening the cost of the standard product.

Wilkinson have so far manufactured 600 million Retractor razors.

Market research

Colour research was carried out alongside name research. The first stage involved researching product names. The next stage consisted of testing

Right: Strong colour, surface effect and form are successfully united in this plastic FADA radio from the 1930s

Below: Red is appropriate for a fire engine, as the colour is instinctively related to fire

the selected names and exploring three colours: red, green and neutral (stone). The comprehensive tests that were carried out involved pack mock-ups, using permutations of the three razor colours and various pack designs. A quantified study was carried out, using a hall test location with 178 male and 89 female buyers from Woking, Ilford and Sydenham. The results revealed that neutral and red were thought of as the most suitable colours, with green clearly 'in the wrong'. Red obtained the fewest negative statements, and was thought to be the most distinctive colour.

Kenneth is aware that it may be necessary to rethink the colour in the future. Companies must always respond to what is happening in the market place. If the shops were suddenly to sell only red products, the colour would have to be changed; the product must attract attention at the point of sale and colour is a powerful tool.

4 Jaguar Cars

Automotives

Colour plays a dominant role in the luxury end of the automotive industry: when economics is no longer the prime motivation, aesthetics and image become the major objectives. Also, the luxury car industry is in a proactive situation; it creates a product that the consumer aspires to purchase rather than supplying a design that is predominantly dependent on market research. It is interesting, therefore, to observe colour in a slightly more elevated position, as well as relating it to the 'hard' exterior and the 'soft' interior in one product.

Background

Ken Rees, principal engineer (colour and trim) at Jaguar, has worked in the motor industry since 1974. He joined Austin Morris to work on colour and trim for Austin Morris MG, Rover and Triumph. In 1979 Ken moved across to Jaguar Cars Limited (a company wholly owned by Ford Motor Company) where he is currently responsible for the colour and material design of all Jaguar's cars.

Ken has been greatly influenced by Brunel, who illustrated that engineering at its purest form is art; McClaren, of the Film Board of Canada, who showed that art and skill should be fun; and John Stark, now of the Rover Group, with whom he worked for the first five years of his career in the motor industry.

The role of the principal engineer (colour and trim) at Jaguar is to determine which colours and materials should be replaced, and within what timescale this should take place. Ken creates the colour concepts and is responsible for these concepts throughout their development until after the production volume has been made. In order to create these colour concepts there are four main issues to consider:

- Knowledge of the consumer and his or her aspirations: this entails more than simply knowing the age and lifestyle of the consumer. The car is viewed not only as a mode of transport but as an accessory or adornment. Through their choice of car – the size, image and colour – the consumer is projecting his or her self-image, as well as requesting that others view and accept this image. If everyone bought a car purely as a means of transport, annual sales of Citroën 2CV or the Reliant Robin would be in millions.
- Anticipation of colour trends through observation, instinct, knowledge gleaned from an ever-changing library of colour reference and trend information, visiting exhibitions, and a consolidation of ideas through being a member of the Colour Group.
- Awareness of technical and environmental risks and opportunities: these facts are presented in research data, resource documentation, trade periodicals and in-house research.
- Styling reviews: these require a consolidation of ideas, and a conviction that they are the correct proposals, in order to convince others of the validity of the concepts.

Ken has a close working relationship with the manufacturing department at Jaguar; initially with regard to the technical development of the paint and later with regard to the paint's performance in production.

The styling review

The product development team helps to organize the styling review at which the new colour proposals are put to the senior directors of the company. Each of the model lines is shown in the proposed new colour, for example XJS and XJ-40; cars are also shown in the colours proposed for deletion. Full background information is provided at the review, including manufacturing feasibility, test results, used volumes and costs. The review takes place approximately 12 months before the main production – 34 months after the original colour was conceived. Jaguar Cars run 17 body colours and eight trim colours and the review committee can accept or reject any of the proposals.

Car interiors tend to change less frequently than body colours, and Ken will usually decide on the time and form of replacement of the new trim colour. Discussions with the marketing and product development teams begin at an early stage, but there is no particular involvement with the manufacturing department unless impact on parts is being taken into consideration. Initially, a single car seat is presented at the styling review, to show detail, followed by a full car set for review. The same procedure applies to change of fabric, with the involvement of manufacturing in this case.

Body colours are designed to create an impact; they exist on their own, whereas trim colours are created to fit within the colour range and work with a number of body colours. Ken can afford to focus the trims on a smaller number of body colours to achieve sophisticated combinations. This trend is currently in development at Jaguar.

Body colours

The body colours are created 46 months ahead of production and ten colours are designed each year. As a starting point, Ken assembles a large board of panels showing all the colours in development, lined up in year order. This board acts as a giant sketch-pad and from here the process of elimination takes place. Ken then spends time (usually about a week) looking at the pattern of development. He can see at this point which colours have recently been changed and which areas need strengthening. Notes are made on the board of panels as it begins to gel into descriptions of colours that will be developed for the next year.

A combination of intuition and expertise allows Ken freedom of selection without market research restrictions, because the top range of the luxury market is such a specialized field. There are, however, a number of 'core' colours which are constant, such as metallic blue and silver. These are colours developed in subtle shifts, with the objective of remaining in the same area but making the colour 'contemporary'. In addition to these essential colours, Ken develops shades which explore the limits of colour possibility. In the context of luxury cars this would be the limits of taste, and more a question of challenging the established concepts of what is an 'appropriate' colour. Jaguar do not necessarily use these colours, but their development is never wasted: exploring the limits expands the middle ground and brings it into sharper focus. The intention is to offer a balanced range of colours that will give maximum colour choice, to allow shades in an important colour area to be included. Additionally, the challenge when met by the paint manufacturers stretches the boundaries of possibilities in other colour-related areas. Ken does not change the concept of the selected colours until acceptance or rejection at the styling review; he feels that retrospective change is the route to chaos.

The brief is passed to Ben Steekers, colour stylist at the paint and paint technology company, Du Pont in Brussels. Ben's involvement is most important: he interprets the brief into paint chips, so his understanding of the designer's feel for colour is vital. (A designer in the motor industry would normally visit paint companies and choose colours already in existence.) Ken works 46 months ahead because all the shades he creates have to be extensively tested in the laboratory and paint shop, as well as being field tested on sites in Arizona, Florida and Australia. These weather tests are based on approximately two years' exposure, and the sites are chosen for their extreme weather conditions.

Jaguar use paint containing a low amount of solids (about 15 per cent solids in solvent) to achieve the best appearance. Medium and low solids contain smaller volumes of solvent and give a duller face.

The body colour consists of clear lacquer over base metallic and solid colours, with low solids technology of around 15 per cent. The base metallics are either aluminium or pure mica (a mineral found as glittering scales); Jaguar do not use aluminium mica mixtures. Micas can be reflecting or refracting, and, if refracting, can be selected to refract at a particular wavelength to provide any required colour. If, for example, a mica is used on a white car, certain angles of the car become bright red, or indeed any desired colour. This effect can be created over any ground colour and it is now possible to create a shot-silk look or a mother-of-pearl effect.

It is vital to bear in mind the body shapes of the cars which will receive the colours, and the light in which they will be viewed. For instance, red-tone blues (dark blues) look fine in the UK, but can appear purple in the USA; and tan works well in the UK but can look orange in the USA. Jaguar once ran a combination of red-tone dark blue with a tan trim that went to the USA. Some very strange telephone conversations followed until they worked out what had happened, although in Europe the colours were extremely successful.

Trim colours

The minimum lead time for trim colours is two and a half years ahead of production, mainly due to the complexity, in manufacturing terms, of introducing change. This schedule includes issuing papers to inform the company of the change, allowing time for the suppliers to produce the new coloured item (this includes testing and sampling for fit and dimension), laboratory testing and field testing. The production line (or track) is never put at risk; there must be full confidence that parts can be supplied and production maintained; and service spares must also be supplied.

Colours come from a variety of sources but always with a specific goal in mind – a model change or date on which they can be introduced. In the past, where the change has to be incorporated into the existing range, Jaguar have tended to introduce changes only to seat colour or carpet, mainly for ease of manufacture. Now, however, Ken designs co-ordinating colours for the whole interior package, which includes seats, carpets, fascia, seat belts and headliner. This approach creates more problems for the releasing system (which controls the introduction of the hundreds of new items involved) and for the manufacturing process (which has a large package of changes to cope with), but the product gains more from an overall transformation than from piecemeal changes.

When Jaguar changed the fascia from black to co-ordinated colour, the effects on parts count was severe. When it was black they had three different levels of made-up fascia (the specifications of instruments and the quality of the wood make the difference) plus left- and right-hand drive. These must all be scheduled so that they are produced to coincide with the correct body when it reaches the track. The fascia must then be stored at the track side. For black, six separate fascias were manufactured, but when coloured fascias were introduced there were six specification levels and eight colours – a total of 48 variations.

The interior leather is coated in waterborne colour. The fabric is 75 per cent wool and 25 per cent nylon and these fabrics must withstand 100,000 rubs on the Martindale abrasion tester after exposure to ultraviolet rays under polythene.

The carpets are Wilton, either polyester or wool. All upholstery plastics are fabric backed, and the rigid and sem-rigid ABS (acrylobutradine styrene) plastic is unsupported expanded vinyl plus a variety of engineer-ing plastics, including glass-filled plastic and unsupported expanded PVC (polyvinyl chloride). All the materials, including foam, are fire resistant.

Problem colours

It is difficult to obtain a clean, crisp colour and still maintain opacity when dealing with dark colours. This is normally achieved by packing the colour with pigments that have the effect of making the colour 'muddy'; this becomes a difficult balancing act and calls for close and sensitive co-operation between the paint company (Carrs), Ken and manufacturing.

Bright colours

These tend to be made up with few pigments to achieve a clean, bright appearance. In some cases, bright colours have been produced with a single pigment and a mica. This, however, causes problems, as batch-to-batch variation cannot then be controlled. There must be at least three pigments to give some latitude for tinting. The solution is to maintain the brightness, achieve good opacity and give the paint manufacturer leeway to tint, in order to compensate for batch-to-batch production variations.

Pale colours

Colours like silver, pale metallic beiges and sands, and light blue are delicate but act as the best inspection colours; they show every speck of dust in a slightly dirty system. The presence of dust leads to a high failure rate on first-time colour inspection.

White

Problems with using white tend to occur during reparation; if a white car is repaired, the masking tape tends to leave yellow stains from sulphur in the adhesive on the tape. Repairs often look 'fuller' than the original colour due to the attainment of full colour strength, as repair adds colour to coat thickness.

Camouflage

Jaguar utilize the properties of specific colours as part of the research and development tests: fluorescent red and yellow on a matt background are used to 'destroy' the shape in order that a 'secret' or pre-launch car can be run in disguise without any add-on camouflage. This is important as camouflage destroys the aerodynamic effect.

All the cars are crash tested at 30–35 mph into a concrete block. The performance of each car is then assessed from high-speed film of the event. Jaguar use bright blue emulsion paint on the cars because it is the most visible colour on high-speed film.

Technical factors

Body

Body colours can be susceptible to failure on the light-fastness tests. The manufacturer must be able to match the original colour on the concept panel, and the maintenance department at Jaguar need to carry out repairs without the repair showing. Colour should not change when it is pumped round the paint system and a great deal of effort has gone into making aluminium particles that can resist damage when going through the process. Micas are very abrasive and cause rapid wear in the paint lines, pumps and spray-gun nozzle, so these all have to be made from stainless steel. Micas also tend to clog conventional paint filters, so the filters in the paint lines are special vibrating units.

Trim

Durability tests are generally similar throughout the car industry but some manufacturers will use paint colours without field testing for durability – they will accept the paint manufacturer's research. Trim test standards vary: there are at least five major test machines for abrasion including Martindale and Taber. There is no cross-reference between any of these machines and the test requirement for the Martindale varies with a low of 30,000–40,000 rubs.

Leather and trim must pass light-fastness-of-wear tests and flammability tests, and leather is also subjected to flex tests. Fabric must withstand 100,000 rubs on the Martindale abrasion tester after exposure to ultraviolet rays. All the materials used in the manufacture of Jaguar car interiors are tested in the laboratory in Canada and, subsequently, at the road-test station in Phoenix, Arizona.

Environmental paint application

Environmental considerations cause new problems for colour application. Waterborne paint gives a better appearance than the medium or high colours (though not as good as the low solids) and almost all shades are available in water. Although investment is needed for any change of technology (and this is particularly so for waterborne paint), this would involve an outlay of around £300 million for colour booth changes including the change from ferrous to stainless steel paint lines. The idea is that water takes the pollution source to the power station, creating large quantities of water contaminated by paint.

It is arguably more sensible to reclaim solvent with large filters of carbon and then recycle it. The solvent is drawn from the atmosphere by charcoal filters and released from the filters by heating. The reclaimed solvent is, at this point, low-grade quality and cannot be used for paint, although it should be possible to enhance the quality by reprocessing. It would cost around £30 million to install the equipment, which is feasible and possibly more environmentally sound. Paint companies are trying all these areas and encountering a large resource problem, while Jaguar are investigating both water and reclaim solvent.

5 Rimmel International

Cosmetics

The role of the designer for cosmetics is two-fold: as chemist and industrial product designer. Cosmetics are a calculated fashion accessory, a combination of chemistry and fashion-related colour specification aimed at a known market. The colour brief for the packaging is to enhance and complement a particular cosmetic while retaining a brand identity for the entire cosmetic range.

The study of Anthropology, specializing in 'religion, ritual and symbolism' inspired Angela Gillic (senior product development manager) to take an academic interest in cultural subtexts of different societies in 'the language of colour'. Through extensive travel in India and North Africa, Angela discovered the sociological and environmental power and impact of colour. Her travels also revealed that colour is not absolute but relative, as seen in the different quality of colour in tropical light.

Beauty on a budget

Rimmel International Limited and Sensiq are part of the Unilever Group. Rimmel are the top volume-producers of cosmetics in the UK and their strength is in the mass market/budget sector which is concerned about 'value for money'. The brand is bought by all age groups within this socio-economic sector. The consumer's perception of the brand is that it offers her the widest choice of colours and products at an affordable price. Rimmel have therefore deliberately set out to create a broad shade spectrum to satisfy the range of demand within their market: for example, they need to offer avant-garde colours to attract the 'experimental' teenage consumer, as well as classic colours to appeal to the more conservative or older consumer. Angela's job is to ensure that the products not only interest teenagers, but in colour terms remain safe enough to appeal to middle-aged and elderly women.

Cosmetics are felt to consist of an infinite palette, where colour choice is dependent solely on the designer's selection or fashion whim. In fact, the cosmetics industry has a restricted colour palette, as only a certain number of colours suit the human complexion, regardless of race.

In terms of strategy, Rimmel are content in the position of 'the first to follow' the innovators working at the premium end of the market. Rimmel imitate innovative ideas such as matt shadow surface textures and 'treatment' mascaras, and launch them six months to a year later at an affordable price. If another company's innovations fail (such as nail-polish pens and multicolour-lead pencils) Rimmel is able to avoid making mistakes.

Through research, Rimmel have discovered that there are distinct differences in colour and texture preferences between different socio-economic groups, and even different religions: the middle-class woman prefers softer, co-ordinated shades with a muted/cream texture; the working-class teenager likes bright primary colours with a high pearl/sparkle finish; and the trendy fashion-victim may want ultra-matt textures in shades that co-ordinate with the season's fashions.

It is very important to be aware of the motivation and aspirations that women have regarding colour in their choice of cosmetics at different phases of their lives. Rimmel use statistical categorization for this purpose. For example: 'The Experimentor' is young and fashion-conscious with a high disposable income. She is adventurous and uses colour to rebel, shock and outrage. 'The Disinterested Rejector' is often a young married mother whose attentions are focused largely on the demands of her children. She may lose interest in her clothes and make-up as she adopts the role of 'Caring Mother'. The woman in this category avoids strong colour statements so she can blend into the background. One of the most interesting aspects of the Rimmel market research, however, is the fact that such cliché-ridden categories still reveal themselves.

Each cosmetic brand creates a colour spectrum to appeal to the psychology of its target consumer. Yves Saint Laurent or Chanel offer a very sophisticated shade range and a 'look' that changes with seasonal fashion trends, whereas The Body Shop keeps its cosmetic shade range muted and natural, and not season-dependent, in line with its ecologically sound marketing strategy.

Product packaging

In the 1950s, when Rimmel was launched on a 'Beauty on a Budget' promotional theme, the packaging was predominantly bold turquoise and white – typical of that period. There was no consistency in packaging design; square boxes sat alongside round or triangular bottles, and there were pointed, flat or curved bottle caps in smooth or textured finishes. As each new product was launched it was designed on an ad hoc basis, so as the range expanded throughout the 1960s and 1970s there was progressively less visual consistency across the products, giving a confusing message to the consumer at the point of sale.

By the 1980s, manufacturing companies were becoming increasingly design conscious and aware of the importance of good, consistent packaging design in encouraging consumer loyalty. Colour is a key element in this design process: baby pink, white, black, burgundy and turquoise were evaluated by Rimmel as potential basic packaging colours but ultimately burgundy was chosen. Market research revealed that burgundy had the most positive attributes; it was seen as being a high-quality, classic, approachable and contemporary colour.

Much time was spent on selecting the exact shade of burgundy, as it can look very hard, cold and slightly purple in some materials. To add some brightness and vibrancy to the colour it was blended with fine self-coloured pearl to create the master colour standard. To allow for maximum flexibility in graphic design, white and gold were chosen as the secondary or accent colours for print and decoration. Gold, in particular, works well for cosmetic packaging as it is perceived as luxurious and glamorous. Over a period of five years, the previous diversity of packaging colours and shapes was replaced by a new, consistent design approach using the burgundy/white/gold colour matrix, a softly angled geometric pack design theme, and a standardized logo and graphics treatment.

Due to the variety of materials (including GPS, HDP, SAN, PVC, PP and PET plastics, lacquer for wood and ink for printing on various surfaces) used by Rimmel for packaging, and the different manufacturing processes associated with each component (such as injection-moulded compacts, extruded tubes and compression-moulded caps) it was difficult to maintain colour consistency relative to the original GPS master. For printing, Rimmel use a pearlized burgundy Vivtor master plaque, in the knowledge that there will be some variation across the range of materials involved. So in the course of product development a new colour standard is set for each component, which is then rigorously quality controlled.

The general trend in cosmetic packaging design in the 1980s, which was mirrored by the evolution of Rimmel's product design, has been to move away from chunky, angular, boxy designs with a severely modernist look and discreet minimal decoration, towards more 'organic' shapes. Softer, rounded curves are replacing hard angles; ovals, egg-shapes, domes and ellipses are taking over from squares, hexagons and cubes. Packs are slim and streamlined, while decoration is becoming more important: textured finishes, speckled effects and the use of different foil-blocking techniques give a softer look to the finished product.

Sensiq use natural forms and finishes to attract the ecologically aware consumer

Just as the cosmetic shade palette for Rimmel has passed from a bright blue/pink/purple colour spectrum into a more subdued brown/beige/peach spectrum, the trend in packaging has turned back to classic neutral colours: black, brown and navy in particular. Fortunately for Rimmel, burgundy is a classic colour that remains enduringly popular with all age and socio-economic groups of target consumers. It is a good colour for show-casing the actual cosmetic shades as it does not clash or dominate; this is particularly relevant for the current neutral shades. Market research reveals that it has now become so associated with Rimmel that any burgundy products are instantly identifiable as belonging to the brand. The colour burgundy is perceived as 'generic' or 'owned' by Rimmel; this protects the range from imitation by competitors and reinforces the positive branding that is essential for the success of any product range in today's crowded market.

Marketing strategy

The point of sale and display units need to be designed in a colour-way that will create an attractive environment to showcase the range in-store, and stand out against competition. Market research shows that different socio-economic groups prefer certain types and colours of packaging; for example, glossy black and gold are preferred by consumers of the lower-cost products because matt textures and neutral colours are perceived as dull, boring and poor quality; a fashion-conscious urban dweller, however, would perceive neutral colours as the height of sophistication.

Rimmel can capitalize on this latent, colour-actuated imagery with their pack designs. For example, the speckled grey/mushroom finish, together with oval forms, reflects Sensiq's use of 'non-animal tested hypo-allergenic/fragrance-free' cosmetic ingredients by evoking a feeling of a smooth pebble, to attract the new wave of ecologically aware consumers, and this has created a lucrative niche in colour cosmetics.

Colour research is carried out 6–18 months in advance of production. Research includes: subscribing to various fashion/colour forecasting services such as Nigel French, Promostyl, International Colour Authority and Nelly Rodi, and textile colour prediction services such as the International Wool Secretariat; visits to Premiere Vision, Indigo, Interstoff and Ideacomo exhibitions; the presentation of colour predictions by suppliers, such as pencil manufacturers like Schwan-Stabilo, Atlas and Pencos;

buying modern and diverse international fashion magazines; membership and participation in the Colour Group. It is also important to attend CTPA (Cosmetics, Toiletries and Perfumery Association) meetings – especially to keep up with EC legislation on the use of raw materials both for packaging (PVC) and colour (cadmium-based colours are now banned). The principal area of contact is with the research and development laboratory at the factory, as it physically creates the 'recipes' for the colours specified by Angela Gillic.

Cascade analysis of sales by product and shade is carried out every quarter by computer. There is also a swatch book of colours containing year-to-date sales, which is monitored bi-monthly. Rimmel review seasonal fashion colours in clothes and accessories to confirm the accuracy of cosmetic colour predictions, while the salesforce collates data on sales trends in stores, of the competitors' colours and products. EPOS (Electronic Point of Sale) and bar coding provide very accurate sales figures at SKU (Stock Keeping Unit) level. Ad hoc market research, both qualitative and quantative, is carried out by talking to consumers about preferences and dislikes of Rimmel and other brands, and focuses on different categories such as nail polish or eye shadow.

Third-party sourcing

The majority of Rimmel's cosmetics are developed in the UK on site, but a certain amount are sourced externally.

- Pencils: these involve a specialized production process due to the raw materials involved. They are sourced at long-established companies in Germany and Italy with origins in stationery or artists' materials, for example Schwan-Stabilo.

Rimmel blend waxes, oils and colour pigments to make their lip gloss

- Kits: these are generally multicolour or multi-product kits, sold as gift packs for Christmas, from Far East suppliers, especially Korea. These are highly labour-intensive and it is much cheaper for UK companies to buy the kits complete. However, many quality problems arise as the powders are often micro-biologically contaminated and need to be irradiated before resale; this process may affect decoration finish. Also, colour choice is limited by non-use of FDA (Federal Drug Association) forbidden pigments that are not saleable in the USA, but are still liked in Europe. This makes the colour palettes subdued and flat, and lacking in the pink/red/violet spectrum.
- Italian manufacturers are famous for their innovative methods of presenting colours. For instance trompe-l'oeil effects are achieved by using duo-chrome pearl sprays to create a two-tone two-dimensional surface effect on a powder. They also create surface interest by pressing several colours together to create a mosaic or geometric pattern. Rimmel import these powders, as the company does not possess the appropriate technology. This is also true of the new formulation types such as lipsticks made with powder and face powders with cream textures. However, technology in Europe is still ten years behind the Japanese in the formulation and presentation of cosmetic colour.

Ingredients

Nail polish

This is basically a nitro-cellulose resin/solution (similar to car paint) with solvents to give the drying film. It is highly flammable, so a separate manufacturing process is required. Additives like nylon or protein claim 'strengthening' properties. Pigments and pearls make up five per cent of the total volume.

Pantone print and fashion colour selections; Colour Dimensions NCS index; RAL swatches

Eye shadow

This consists largely of talc, mixed with an oily mixture called 'binder' to facilitate pressing. The texture of the talc is crucial, so repeated milling gives a fine, sheer texture. Talc can be siliconized for smooth application. Pigments and pearls make up 40 per cent of the colour.

Lipstick

Lipstick is a blend of waxes and oils including natural or synthetic caster oil. Lanolin derivatives give moisturizing properties. Colour pigments and pearls determine shade (15 per cent volume of the formula). Most lipsticks are fragranced.

Mascaras

These are water-based emulsions consisting of waxes, film former and pigments with or without fibres for lash build-up. Solvent-based emulsions

Glossary

ABS	Acrylo butradine styrene
GPS	General purpose styrene
HDP	High density polythene
LDPE	Low density polythene
PET	Polythylene terephthalate
Phenolic	Compression moulding compound
PP	Polypropylene
PVC	Polyvinyl chloride
SAN	Styrene acrylo nitrol

consist of solvents and waxes with colour pigments. They are water resistant and have a strong smell. Colour for both products amounts to ten per cent.

The issue of testing colours and formulas on animals is an emotive topic of interest to consumers. One point of view proposes that it is inhumane to 'torture animals for the sake of beauty'; another that it is not safe to put untested cosmetics on a face. Whatever the view, this debate has led to tighter controls on the cosmetic industry.

6 Newell and Sorrell

Corporate identity

Frances Newell trained as a fine artist and then as a designer. In the early 1970s she worked primarily as an illustrator, before founding Newell and Sorrell with John Sorrell in 1976. Newell and Sorrell has gradually grown to become one of the UK's leading design consultancies, working with clients such as InterCity, Jazz FM, Parcelforce, Waitrose, The Boots Company, Berol and The Body Shop.

Frances believes that the use of colour in design can communicate many different ideas, emotions and messages: 'Colour can reflect what an organization is all about and send signals about its personality.'

Every organization has many opportunities to use colour positively. Reception areas, canteens and restaurants, meeting rooms, offices and even lavatories are all spaces in which colour can help to welcome, relieve stress and say something good about the organization, if it is chosen with care and creativity. The use of colour on the signs, stationery, brochures, reports and other two-dimensional representations of an organization can help to clarify, inform, communicate and create a favourable impression. Vehicles provide a huge canvas for the use of colour and when it comes to workwear, the colour must not only relate to the organization but take into account the requirements of the wearer.

If chosen successfully, colour can help people both inside and outside an organization feel good about its environment, its product and the service it offers. When Jazz FM, the UK's first radio station dedicated to jazz, asked Newell and Sorrell to create an identity for them, it became clear to

Karisma pencils by Berol

Frances, 'that the wild, crazy, spontaneous, moody and random characteristics of different types of jazz – blues, New Orleans, big band, soul and so on – could be visually suggested with an equally variable and exciting colour palette'. The Jazz FM identity was launched using a brilliant purple and orange colour scheme, but a wide range of other colours was soon introduced for the logotype and other elements of the identity. To demonstrate the rich potential of the idea, a painting was produced to illustrate over 40 different colourways. Jazz FM went on the air in March 1990, supported by an animated TV campaign, posters and multi-coloured stationery range, to instant and wide acclaim.

The InterCity corporate identity

Frances believes that organizations are usually perceived in the light of their products. She also points out that the use of colour on products and packaging affects a buyer's decision at the point of sale; the consumer may then use the product in the home for a long period of time, so its colour becomes part of the environment.

It is for these reasons, says Frances, that manufacturers carry out extensive market research to test consumer reaction to colours, and much experimental use of colour falls by the wayside. The problem here is that the research often picks up people's responses based on previous experience which may not be relevant to the future.

Colour plays a central part in communicating messages through design, as Frances observes:

> It is extraordinary how finding the 'right' colours for a project can trigger an emotional response that would be impossible to capture in any other way. The 'right' colours, though, are not always obvious. But it is equally important to know when to use a little – to intensify the effect of the overall design. Very often, with colour, less is a great deal more.

Karisma

Karisma pencils by Berol have been chosen as a case study because the product inspired the use of natural colour; the materials chosen are reflected in the colour palette; and the sophisticated confidence in the underplay of colour has produced an original concept. A holistic approach within good design has created an ethos behind the selection of materials, colour and texture that is sympathetic to the ecological movement. This is refreshing to see amid the all too common occurrences of a 'green' afterthought in many areas of design.

Berol can trace their manufacturing history to 1856. Over the decades, pencils have been a core product for Berol but they have been sold almost exclusively to the commercial stationery and educational markets. Traditional coloured pencils were manufactured but these were aimed mainly at the educational market; Berol were only on the periphery of the graphics and art material market in the UK.

In other countries where different Berol affiliates operate (for example, in the USA) Berol have a presence in the graphics market, particularly with the internationally recognized Prismacolor pencil. This product is not available in Europe. To enter the UK and European graphics market, Berol UK had to consider developing new products or repositioning existing products, or both.

In 1986 Newell and Sorrell began to create concepts for a brand that would enable Berol to launch a range of graphic products into the European market.

The brief

The objective was set in the original brief to Newell and Sorrell:

> To enter the graphics market with a fully integrated, branded, visually attractive range of freehand consumable products and materials and to achieve a significant market share in three years.

Because the market was a new one for Berol, a creative strategy was required which would meet the special needs of the market. To achieve that, and to achieve Berol's objectives, Newell and Sorrell established a creative strategy that would be effective in all aspects, from products to packaging (making the packaging itself an essential part of the product) and then to merchandizing and promotion.

The creative strategy had three principles:

- visual excellence
- fitness for purpose
- uniqueness.

The Karisma product range was planned to be extensive, providing products that are used in design studios, drawing offices and art schools and colleges.

The design process

Frances explains that the product had to appeal to the users: these were illustrators, artists, designers, architects, draughtspeople and students, who are all visually aware and interested in how things look, work and feel. She points out that the style and colours of the packaging did not have to be 'strident or garish to be understood' but that detail was an important factor.

The products of the Karisma range were to be diverse; they would perform contrasting functions and be made from different materials. Newell and Sorrell suggested to Berol that the diversities should be reflected in the style and manufacture of the packaging as well as in the products. The aim should be to select materials which were sympathetic to each individual product type as well as being interesting in themselves, for example paper, wood, metal, fabric or plastic.

The first two product ranges incorporated a different type of packaging relating to the product: paper-wrapped boxes were designed for pencils; and sleek vacuum-formed plastic boxes were created for markers.

The first Karisma product to be launched was a range of pencils: this included aquarelle (water soluble) drawing pencils, a range of 72 coloured pencils, and dense blacklead drawing pencils. All of the existing competitive products had a paint finish which approximately matched the colour of the leads, and they were all packaged in bright metal boxes. Newell and Sorrell recommended a quieter, more sensitive approach.

Newell and Sorrell were inspired by the nostalgic quality of pencils and their tradition of everyday use. They were also influenced by the material used: wood is pleasant to look at, to touch and to smell. They decided that the wood casement of the pencil should remain unpainted, in order to display the natural colour and texture of the wood.

The problem then was how to make each pencil identifiable when in use. The solution entailed cutting the end of each pencil, to give a chamfered effect; the lead is exposed here as well as at the tip, enabling the user to see the exact colour of the pencil. This created a distinctive pencil with an appropriate 'art and craft' look.

Much of the inspiration for the packaging and detailing came from looking at traditional artists' equipment and stationery products such as canvas, wood, plain metal containers and eyelets, brown-paper document ties and brass paper-binders. A great deal of material was collected to select colours and textures that would enhance the pencils, and also display the colours of the leads. Black was the most effective colour to highlight the colours of the leads, so it was chosen to line each box and tray.

The outside of the boxes was wrapped in textured, recycled paper: the aquarelle pencils had a soft grey-blue wrapping, while the coloured pencils were wrapped in greenish-buff paper and the blacklead pencils a soft grey. The finishing details included brass eyelets, and on the larger boxes, document ties and brass hinges.

Identifying the products on the outside of the boxes was a problem:

product information was printed on the lids, the atmosphere would be lost and the user would have to endure this, by then redundant, information.

The solution was two-fold. All the vital product information, which needed to be bold, was printed on a separate band of paper wrapped around the lid. This could be removed once the pencils were in use. Newell and Sorrell then devised a series of hallmarks that were embossed directly onto the lid of each box, and onto the pencils, as a statement of quality.

In the first year of the Karisma launch Berol's presence in graphics retail rose from nil to 700 outlets in the UK and the products are now exported to Europe, the USA and Japan.

■ InterCity

The process of colour selection for the British Rail InterCity corporate identity had the restriction of having to relate to colours on already existing locomotives and rolling stock. This case study follows the process of creating corporate identity, involving a defined colour palette, spreading across the design disciplines from packaging and literature to locomotives and carriage interiors.

Newell and Sorrell were first approached by Dr John Prideaux, the director of British Rail's InterCity division, in December 1986. At the time, British Rail consisted of five operating units, one of which was InterCity, and it was Dr Prideaux's task to transform a £125 million operating loss into a profit by April 1989. He based his turnaround strategy on three key elements: quality, efficiency and growth.

The first stage of work consisted of discussions, investigations and analysis which led to the development of initial concepts of the visual identity of InterCity. The basic elements of the identity were developed in order to complete the priority applications; it was important to embark immediately upon the development of the train livery, for example, as this is a highly visible element. Frances Newell explains:

> At the time we were first approached to produce the corporate identity, InterCity was presented in a very downbeat way. The logotype was in the same visual language as the signs telling people where the toilets were, and InterCity came over simply as part of an information system. It did not appeal to people's emotions. It had no personality.

To produce a successful corporate identity, it was important to clarify the nature of InterCity, its function and its future plans. Finally, the image of a swallow was developed as an emblem for InterCity; the design reflected the warm and friendly aspects of the organization. It was decided that an abstract symbol could not convey this message; in the same way, the image of a cheetah was considered but then rejected as too aggressive (even though it suggested speed and energy).

The final creation was a visual system comprising a number of key elements that can be used in various applications: the colour palette, the InterCity logo, subsidiary brand logos, subsidiary typefaces, the red track, the use of the British Rail symbol and the emblem. All of these elements converged to present InterCity as the flagship of British Rail.

The colour palette

Newell and Sorrell were aware that the chosen colours would have to work with the existing locomotives and rolling stock for some years ahead. Their choice of colours was strongly influenced by the trains themselves, which were painted principally in two colours: a light buff on the lower half of the carriages; and on the upper half, a special colour that InterCity had developed for its ability to conceal the inevitable dirt and spray from the track. This colour was a very dark, brownish black, and was nicknamed 'moleskin' by British Rail staff.

Bright colours were not used, as Network Southeast had repainted all of its trains in a strong blue, red and white scheme; InterCity, as British Rail's flagship division needed a contrasting appearance. Using the 'moleskin' as a base, a strong monotone palette was created, relating to the engineering atmosphere of the trains.

Silver was selected for the swallow emblem, appearing on the locomotives with a highly reflective finish. The logotype also appears in silver matt, and both of these work well against the dark 'moleskin' on the locomotives. White, pale grey and black were added to the base palette, together with the British Rail colour red, which is only used occasionally as an accent. This controlled colour palette has worked successfully across applications such as literature, tickets, timetables, maps and menus.

The monotone colour scheme was retained for the new coach interiors for the InterCity 225 with the addition of mid-greys for the first-class coaches. By using a very bold, contrasting design for the carpets and polished metal silver handrails, life, energy and sparkle were added. The passengers then complete the picture by adding colour. The colour palette is still being expanded: subtle variations are being introduced into the grey range, while rich dark reds are being tried as an accent.

Because InterCity is so large, the range of applications for the colour palette was vast. A book of guidelines for the use of the InterCity identity was produced, including the referencing of the InterCity colours and their usage, as well as a range of typefaces that can be used in various situations.

InterCity has now turned its operating loss into profit: although the corporate identity cannot take all the credit, it has played a significant part, both practically and psychologically. The new identity is continually applied to all aspects of the organization, including new trains, corporate and promotional literature, uniforms, environments and packaging.

7 Scantest

Colour market research

Market research has become a key factor in colour selection. The way information is obtained can affect the results of a market research programme: variants created by factors such as geographical areas, time of day or question order can give rise to varying interpretations. It is necessary, therefore, for designers, manufacturers and clients to contribute to market research programmes.

Research methodology

Bill Dunning is managing director of Scantest Ltd, a company specializing in evaluating the market potential of new designs and colours on an international basis. In 1974 he developed the Scantest technique for forecasting the future performance of new designs and colours. Tricia Dunning is both a director and a design and colour consultant at Scantest Ltd; she trained and worked as a fashion designer and illustrator before becoming involved in Scantest in the mid-1970s. Scantest Ltd (part of the Scantel group) have associates in North America, Continental Europe and the Far East, and they specialize in design and colour research in various industries.

The company is built around a single methodology – the Scantest technique. This is used for testing anything that includes a significant aesthetic element: the interaction of colour, shape, design or texture. Scantest interview the public and demonstrate products; they then gauge and record the reactions. These are then fed into a computer programme which reveals a rank order from the best to worst seller. Scantest and the marketing team assimilate the information and make it available to the designer; this ensures that the designer does not waste time designing for the wrong segment of the market.

Scantest was created with the purpose of reducing the subjectivity in decision-making, thereby minimizing the risk of targeting errors and associated financial penalties. They aim to provide an accurate evaluation of the market for any given colour; to optimize colour range composition for any given number of colours; and to aid production and stock control. They also help to discover why some colours are more successful than others, thereby facilitating new product development.

The interview

The Scantest technique follows a strict interview format which is constant, to allow for reliable and reasonable results. The interview group can be a cross-section of the general public or, more frequently, a sample of target buyers for the products being researched, and interviews are always split between northern and southern locations. Initial interviews are carried out in a busy high street or shopping centre, and interviewees that fit the recruitment criteria are then invited to a nearby hall or studio to view a number of test colours.

The number of people interviewed depends on the amount of colours tested: an average test of 20 colours would require a sample of 200 interviewees for a reliable result. Most of the information collected is numerate; this can be entered into the computer programme to provide results which give a straightforward ranking of consumer appeal and a percentage score for each colour tested.

The results can facilitate decisions on which colours to launch and which to reject, and the computer programme can provide forecasts of sales for any number and combinations of colours that clients wish to consider prior to reaching final colour decisions.

The interview format

1. Different colours of stimulus products/material are displayed to respondents.
2. All the colour options are viewed before questions are asked.
3. Respondents are taken to each colour and asked (using a prompt card) how much the colour is liked or disliked on an aesthetic level.
4. Colours which do not score positively on aesthetics are removed from the range of colours.
   (NB Order-effect bias is minimized by listing the colours in rotation.)
5. All colours which have scored positively on aesthetics (known as the 'Hollywood Dimension') are re-examined, and respondents are questioned about their suitability for a particular situation. Colours which do not receive positive scores at this applied level are removed from the range of colours.
6. Respondents are asked to review the 'successful' colours and provide a ranking order.
7. Reasons for selecting a particular colour as first choice can be elicited to provide diagnostic insight.
8. Propensity to purchase the product in the first-choice colour is recorded.
9. Colours are 'profiled' on a series of image statements or attributes, for example, 'fashionable', 'modern' or 'expensive looking'. This data provides 'maps' of perceived strengths and weaknesses for each colour.
10. Respondents are invited to look again at the complete range of colours and asked to choose the least liked, giving reasons.

Accuracy

Where Scantest forecasts of popularity have been checked against actual sales results, in the paint market, for example, correlations as high as 0.95 have been achieved.

Qualitative research, which is unstructured (in groups or individually) reveals that younger people have a greater awareness of the role of colour, with a holistic approach. This is largely due to the evolution of the high street: shop windows and room sets are seen to portray colour with an acknowledgement to its importance. Stores like Next, Habitat and Laura Ashley all have an individual style, and present pre-sourced, packaged looks to the consumer.

When working as a colourist for Scantest, Tricia is often required to carry out a projection of how the colour palette is going to move. She observes individual colour families and reports on how fashion is going to influence colour and what is happening in areas such as electrical goods and floor-coverings.

It is important to be aware of the effects that colours have on each other; for example, how orange would affect the general yellow palette. The client must be informed of the effect of choosing, for example, a strong yellow palette: how does it dominate the total palette? Is it brash, acidic or day-glo, or does it give everything a slight sepia wash? Tricia then has to 'bring [the palette] down to a compact unit of colours that are important, not just as main colours but as accent and highlighter colours'. Tricia must also ensure that the client understands that she is not promoting high-fashion colours for a product that has a two-year lead time and might not lend itself to any of the fashion colours.

Fashion colours tend to come and go very rapidly. Brown, for example, has undergone some major status changes as a fashion colour and its popularity has been cyclic. Browns dominated the market at one point, and then everyone was caught on the hop – it was suddenly in decline. This led to the growth of greys. Now kitchens are full of natural woods, and brown has come back by that route.

Colour use in products follows different patterns according to country and Scantest are also aware that there are 'classic' designs that are in a sense pan-European designs. Most European maunfacturers want this core range of products. The client will ask Scantest to find these core designs so that they can identify the design in the various colourways that are idiosyncratic to particular countries. The client still has a core range, and they are still able to identify how many countries will buy their products – they can simply match colour to country.

8 TCS

Furniture

When relating colour to a product, whether working with plastic, metal or fabric, it is the relationship between shape, function, price, market sector and identity that has to be taken into consideration. Although TCS manufacture two different types of product (floor-covering and furniture), the production processes are similar in many ways, and both products are eventually placed within a domestic, contract or industrial interior.

This case study is unusual, as the company is uniquely both the client and the specifier, and the designer is working with external manufacturers throughout the design process.

The client as specifier

As part of the Department of the Environment, TCS sell products and services primarily to government departments and the armed services at home and overseas, but also to other organizations within the public sector, such as British Rail.

It is relevant, when looking at the relationship between colour and design within industry, to observe the role of the designer using colour to create one product and divide it into two separate ranges for clients with disparate needs and attitudes.

Teresa Collins, in her role as textile designer/design manager for TCS, is responsible for both the contract flooring and chair projects. Her job description has expanded over the past few years, as advising on colour and finishes becomes more important. Her role is to develop new ranges, and to improve the design standards of products such as floor-coverings, furniture, upholstery or furnishing fabrics.

The colour work in the factory is initially carried out by the designer and the design director. Colour matching initially takes place with yarn tufts rather than scientific permutations, although the matching is actually executed in the dye laboratory. Teresa produces her design and colour proposals for fabric, paint finishes and floor-coverings, and then takes these through to production alongside colleagues in the relevant sections. She also advises TCS's product and interior design sections on the use of colour and finishes in their work. As a basis for her proposals she sources new developments, and analyses colour and design trends.

Teresa works in two specific areas: the contract/office market and domestic/residential furnishing.

The office environment not only includes workstations, but conference rooms, boardrooms, reception areas, interview rooms, canteens, restaurants and coffee bars. Today the consumer has a wider choice of products as there is open competition with other suppliers, and even in the public sector the traditional rules are less rigidly applied.

In general, whatever the current trend, contract colours tend to be stronger, more definite shades. Colour must be applied in a practical manner; people do not want their environment to look shabby or dirty, but they also need to feel comfortable. The integration of lighting and office equipment requires important considerations; strong, saturated colours absorb light and can look oppressive, yet if there are too many pale surfaces, they can cause excessive light reflectancy leading to headaches and eyestrain. It is crucial to strike the correct balance.

Residential accommodation must also be extremely practical, as a number of different families will live in a house over a ten-year period. The schemes are deliberately neutral, so that colour can be introduced by the individual. These interiors also need to withstand heavy use, with continual day-to-day wear and tear.

The difference between the contract and domestic markets lies in the balance between status and function as perceived by the consumer. This is reflected in the amount of money spent and the attitude towards spending that money. For example, when designing for a prestigious office suite, or fine historic residence, status and appearance are of the utmost importance. This is reflected in the colours and materials chosen for such interiors: they are rich saturated colours, plush carpets or wooden floors – high-quality finishes that suggest wealth.

There is an attitude change when designing schemes for general working environments and living accommodation. Function becomes the key priority, so that saturated colour is often shied away from, leading to a preference for grey, neutrals and just a suggestion of colour. These shades can look bland if they are not carefully put together; an approach known as a 'sea of brown and beige'.

With a low-budget scheme, it is important to avoid creating an 'institutional' environment, whether drab or 'cheap and cheerful'. In the past, walls were often dull, mid-tone colours, and very bright 'highlight' colours were introduced as a contrast. This application of colour is now considered very dated, and designers are made aware that it is always important to bear in mind the effect these colour schemes can have on the people who live and work in this environment.

Research

Colour research is vital. Through personal observation and the visual research team, the designer is aware of what is happening in the high street. Noting articles in architectural, design and fashion journals, looking at advertisements, exchanging ideas with other designers, and visiting trade fairs, manufacturers, museums and exhibitions are relevant to colour research. It is important to be aware that new colours should work within existing ranges of sister products, otherwise the new range will look alien. Although, this may be a 'safe' option it must still have something original to offer, such as a range of colours put together in a commercial but unique way. It is important that a range of colours is offered as a group, ensuring that the colours work for the entire range. Finally, the consumer must be offered a real colour choice. It is psychologically important that consumers feel they have a decision to make, and that they have a good selection to choose from and not just 'any old grey' in ten variations.

The chair

Independent market research, reviewing the whole of TCS seating, identified a need to introduce a range of 'up-market' general seating. This was to be suitable for committee and conference use.

This case study looks at one of three chairs that have reached production stage. It is the 'up-market steel-frame chair', chosen because it incorporates several finishes and it relates to a table that was produced to complete the furniture needed for a committee workstation.

A project brief was drawn up to specify constructional and functional requirements. At this stage, design, purchasing and marketing were all involved. The brief stated that the basic construction should be metal, available with and without arms, and upholstered in a range of fabrics with optional colours and finishes. (Offering a good range of colours with optional finishes increases consumer appeal.) The chair would have to pass specific structural and fire tests. Price guidelines were incorporated so that when the chair was marketed it would become highly competitive.

Carpet tufting from Chromatone; Merck iridescent plastic swatches; Colorworks Vol 1

There are three individually treated components to the chair: a metal framework, seat and back pads, and legs. The metal framework is the constant factor, due to the high volume produced in manufacture. Selecting the colour and finish for this component was of primary importance: a hard-wearing paint finish was needed, and as the arm of a chair is a tactile area, the designer felt that a slight texture would create the right surface, therefore a 'rextel' paint was specified. This has a modified texture that gives a warm tactile quality when sprayed onto the metal.

In contrast to the warmth of touch a cool blue-grey was chosen for the colour. This colour has a chameleon-like quality, in that it subtly changes colour when juxtaposed with rich bold fabrics.

The seat and back pads are upholstered, and by offering a good selection of fabrics (plain, textured and patterned), in a range of colours, the element of flexibility can be introduced. (This range also included the option of customer-selected fabrics.) The materials were chosen for their sophisticated colours, and a subtle, textured 'plain' range was compiled that would complement the paint finish on the arm. A two-coloured random-pattern weave was added to the range to create an area of visual interest.

A variety of finishes was also incorporated into the design of the legs; these were primarily natural beech (with the option of being stained or painted) or chrome.

The colour and gloss finish of the paint had to be considered. Various gloss levels are available ranging from dead matt to full gloss; there are nine available options. Although gloss is more hardwearing than matt, Teresa felt that a full gloss looked too severe, so a 70 per cent satin was chosen. To contrast with the natural beech, a warm, rich mahogany colour was chosen to be offered as an alternative, giving the chair a different 'look'.

This was thought by some consumers to look more exclusive and expensive; a prime example of 'preconceived ideas' of quality.

The resulting chair has been widely acclaimed. The clients were impressed to see a new chair that is both functional and visually pleasing. Architects actually found the concept most appealing as they could, by the nature of their work, visualize the different looks that could be achieved.

9 Giant

Packaging and graphics

Alan Herron is a graphic designer and one of four founding partners of Giant, a small graphic design consultancy. They design and produce packaging, corporate identities, annual reports, exhibitions and literature of all kinds and their clients include the Arts Council, The Boots Company, Q8 Petroleum, Woolworths and W H Smith.

Colour influences

Everything Alan sees influences the shape, form, texture and colour of the projects he designs. He is greatly influenced by television, fashion, cinema, nature, sport, magazines, books, architecture and technology. Alan feels that he can store images rather than facts in his memory, and that he can recall them with greater ease; for instance, he can remember seeing a yellow crane standing in front of a blue warehouse, and storing the colour relationships. This has led him to interpret the accidental nature of how colours juxtapose with one another in his work, by using colour combinations that would not normally appear together. He is also far more aware of scale through these 'accidental' images; for example, if he watches a red ball bouncing on a bright-green synthetic football pitch, he sees it in terms of a small red piece of type or detail, against a green package or double-page spread. Alan believes that colours fluctuate according to trends dictated by the high street and ultimately, fashion designers. If he sees too much of a particular colour combination he might feel inclined to avoid using it, whereas, if a certain colour scheme appealed to his sense of humour it may be used. Alan feels colour humour is important if used intelligently within the context of the design.

Alan finds it difficult in many cases to persuade the client to agree to unusual colour combinations. The client normally feels safer with tried and tested combinations like navy blue and red, primary colours or green and yellow. The client may indeed wear sophisticated shades in the shape of a sage-green tweed jacket, or unusual colour combination ties, but if these colours are suggested for the cover of an annual report they become nervous and

Early development work by Giant for KPL products

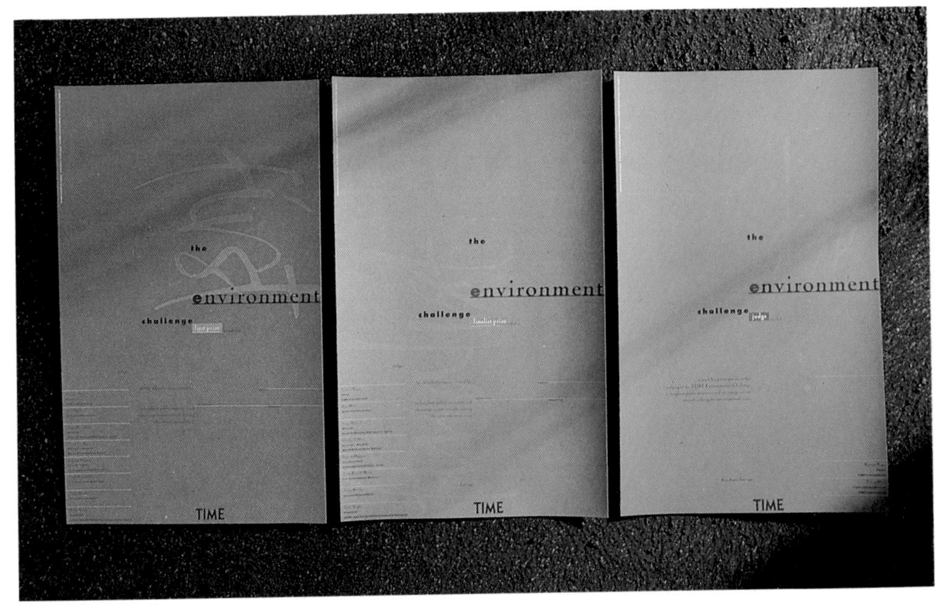

Subtly-coloured award certificates created by Giant to replace the conventional white certificate

conservative. Colour is tremendously under-valued as a subliminal communication tool, and the client is often wary of new colours because of lack of understanding. They often see colours as having a particular role: navy blue for the boardroom, brown in a farmhouse kitchen or yellow in the form of a JCB tractor. However, Alan finds that if the colours are selected carefully and intermixed with other shades, their personalities change and give different messages.

The lifespan of much of the work done in graphics is very short; this usually means that the use of colour can be much more adventurous in graphic design than in product design. Many projects such as annual reports, packaging and literature need to have impact and be eye-catching; in this situation colour comes to the forefront.

It can also be said that the work of the graphic designer is often disposable – once the information has been passed on the graphics can be discarded. Products, however, involve a much higher capital investment, so colour cannot be whimsical or transient in this sector. The industrial designer cannot afford to take risks: products take years to change colour because marketeers need to be convinced that consumers will devour vast quantities of pink polka-dot washing machines before they can invest in their manufacture. If a manufacturer has successfully been making black and white products for 40 years, it would take a strong argument to convince them that blue products will be the next market leaders. Alan feels that:

> Perhaps it is up to the consumer to create the demand for products in more adventurous shades. Manufacturing techniques are such that once a product has been purchased in white, say, even though the tastes of the purchasers may change within the lifespan of the product, current technology doesn't allow the product colour to change. Why can't the consumer change the colour? Why can't the consumer spray panels in the colour of his/her choice to slide into a washing machine/fridge? The consumer doesn't seem to have much power in dictating what is manufactured.

Alan observes that an industrial designer designs a product and then proceeds to choose the colour. During the process of colour selection the product designer must consider the fact that the choice of colour will dictate whether the design sells well and that black and white are safe colours and therefore good sellers. It is not until the end of the design process that the product designer makes the crucial colour decision, whereas a graphic designer must make decisions about colour much earlier in the design process. Graphic designers regard colour as an integral and essential part of their work, while to many industrial designers colour is peripheral and additional.

Kuwait Petroleum Lubricants

Colour played a significant part in the solution of the KPL project. (The case study follows the design process, but cannot supply a consumer result as the product was newly launched at the time of writing.) Product design and colour were interrelated in this particular case study, as the image of the product within the container is dependent on the brand identity.

Although colour is the key to the solution it is necessary to look at why Giant were chosen by KPL for this particular project. Alan explains:

> We convinced the client that we could produce a hard-hitting and practical solution to the brief. We had worked with KPL before, they knew us and how we worked. They trust us. It was also important that our price for doing the job was more cost effective than our competitors'.

The brief

KPL are a division of Q8, the petrol retailer. Q8 have several products in the premium-quality product area; KPL had a portfolio of slightly lower-quality products and needed to give them a consistent brand feel without competing with the Q8 products. The new brand would have to be perceived as 'top of the second division'.

Within the overall brand, there were four distinct product areas to be differentiated. The typography had to be clear and legible, while the colour was to be used to attract the customer, without becoming too sophisticated. Colour was used to position the products: if the colours were too tasteful, the customer might mistake the product for a competitor of Q8. The product was to be placed against similarly priced products; in this area of lubricant retailing, price is the all-important criterion.

The graphics had to be applied to the various bottles of motor fluid as

Multicoloured telephone wires illustrate the successful union of colour, form and function

self-adhesive labels. These labels would be litho printed using a maximum of four colours for each product area, while the plastic containers could be made to match any colour. Giant were aware of colour-matching difficulties that could occur between the plastic container and the ink on the label.

Giant presented KPL with several design directions, each using different colours, layouts and typography. Although it is more usual for clients to wish to see variants on a chosen direction (using a different typeface or alternative colours), KPL selected a design direction immediately.

The solution

The chosen direction uses colour boldly to differentiate between the four product areas. Each product is identified by clear typography, as bold typefaces hold a good amount of colour, thereby making the design more visually interesting. The calibrations on the side of each label communicate the idea of science and technology. A dark colour was chosen for the container, as a contrast to the bright label colours. To minimize perceived colour differences the dark container colour was used very sparingly on the printed label. The objective was achieved by interrelating the typography and colours, but not at the expense of clarity, as Alan explains:

> We spent a lot of time and discussion in selecting colours and layout for this project ... the client felt that simplicity and particularly colour would help their product to be perceived in the most appropriate way ... certainly we would say that the final product design is colour-led. We feel that we've successfully combined a relevant marketing solution with an interesting juxtaposition of colour.

10 Conclusion

Guidelines

- Colour should be considered in relation to form and function.
- Colour should be seen in context with texture and surface effect.
- Colour should relate to the materials used.
- Colour should relate to the product's ultimate environment.
- The colour specified must relate to the product, function and market; it should not be chosen simply because it is a 'fashion' colour or because a particular product has always been associated with a certain colour.
- Highlight, accent or trim colours can enhance or co-ordinate the main colours of a product. The product itself can also perform a highlighting function within an environment (for example, a red saucepan can be effective in a white kitchen).
- Colour is an important safety factor; a colour 'code' such as yellow and black warns that a toxic substance is present.
- The fact that it is cheaper to change colour than to redesign, should be considered.
- The light source should be considered; metamerism (the distortion of colour by different types of light) is a serious colour problem.
- Any available sources should be used to predict and observe colour trends, including: art, fashion, music, politics, economics, food, exhibitions, forecast and notation systems, and world issues and events.

- The fact that colour has psychological and physiological effects on human response should be considered.
- Proportional use of colour controls the perceived colour; for example, placing red and yellow together affects the appearance of both colours.
- The economics of industrial design dictate that colour specification must be carefully researched.
- From concept to marketplace, all areas within industry – management, manufacturers, designers and marketing personnel – must exchange ideas about colour selection.
- It must be noted that with the advent of the Single European Market, the EC colour-related standards and restrictions on colour will be the norm in the UK.

Aspects of colour

As an introduction to the importance of colour in industrial design, this book has attempted to provide an insight into the interrelationship of colour and product from concept to marketplace. This has entailed the inclusion of case studies from various industrial design disciplines, and in doing so has shown a similarity in sourcing, market research and colour specification between unrelated products like cosmetics, automotives and housewares. The colour-related attitudes, problems and solutions encountered in the case studies should prove to be helpful when applied to the reader's own design and manufacturing experiences.

In the arena of colour specification it is necessary to consider aspects like colour and light (colour cannot exist without light, and light controls the perceived colour) as well as a number of related topics, such as biology, physics, chemistry, psychology, optics, aesthetics, art, colour blindness (one in ten males is colour blind), colour cycles and lead times. This book has introduced these topics when appropriate but an extensive bibliography is provided for further reading and research (see page 94). The combination of aesthetic, functional and psychological colour choice, product performance, marketing strategies, financial considerations and technical procedures makes colour in industrial design a fascinating and continually expanding area.

A new era

Increasing colour choice in the 1980s has given rise to a new wave in retailing: many stationery shops provide a wider choice of colour, pattern and surface texture for items like the school exercise book, rather than the conventional two or three colours; fashion and cosmetics are sold as palettes of colour, rather than in a few 'fashion' colours; baby products such as pushchairs and highchairs are sold with colour choice as a dominant feature; and the teenage market is acknowledged through products such as the 'Le Clic' camera, produced in purple with a lilac trim, and the Philips 'Tracer' shaver in a range of primary colours.

Children are now computer literate at the age of five or six and they are making colour decisions on computers as part of their school education. This combination of increasing technology-awareness and an everyday acceptance of colour choice provides a background that will perpetuate the notion of placing colour alongside form and function.

These factors, combined with increasing realization in industry that colour has to be dealt with sympathetically, suggest that there will be more emphasis on colour in industrial design in the future.

Bibliography

Albers, J (1975) *Interaction of Colour.* New Haven: Yale University Press.

Babitt, E in Birren, F (Ed) (1967) *The Principles of Light and Color.* New York: Citadel Press.

Beck, J (1972) *Surface Color Perception.* New York: Cornell University Press.

Billmeyer, F W and Saltzman, M (1981) *Principles of Color Technology.* New York: John Wiley & Sons.

Birren, F (1977) *Principles of Color: A Review of Past Traditions and Modern Theories of Color Harmony.* New York: Van Nostrand Reinhold.

Birren, F (1937) *Functional Colour.* New York: Crimson Press.

British Colour Council (1964) *Colour and Lighting in Factories and Offices.* London: BCC

British Colour Council (1951) *The British Colour Council Dictionary of Colour Standards.* London: BCC

Chevreul, M E (1981) *The Principles of Harmony and Contrast of Colors.* New York: Van Nostrand Reinhold.

Clarke, R (1952) *Food Colours.* Leatherhead: British Food Manufacturing Industries Research Association.

Conway, H (Ed) (1987) *Design History: A Student's Handbook.* London: Allen & Unwin.

Danger, E (1969) *How to Use Color to Sell.* Boston: Cahners Publishing Co.

Danger, E (1987) *Colour Handbook: How to Use Colour in Commerce and Industry.* Aldershot: Gower Technical Press.

Ellinger, R G (1980) *Color Structure and Design.* New York: Van Nostrand Reinhold.

Favre, J P and Novembre, A (1979) *Color und/and/et Communication.* Zurich: ABC Editions.

Gerard, R (1957) *The Different Effects of Colored Lights on Physiological Functions.* Ph.D Dissertation: University of California.

Heskett, J (1980) *Industrial Design.* London: Thames & Hudson.

Hunter, R S and Harold, R W (1987) *Measurement of Appearance.* New York: John Wiley & Sons.

Itten, J (1970) *Elements of Color.* New York: Van Nostrand Reinhold.

Itten, J (1974) *The Art of Color.* Van Nostrand Reinhold.

Jones, H (1950) *Planned Packaging.* London: Allen & Unwin.

Judd, D B and Wyszecki, G (1975) *Color in Business, Science and Industry.* New York: John Wiley & Sons.

Kandinsky, W (1947) *Concerning the Spiritual in Art and Painting in Particular, 1912.* New York: Wittenborn Schiltz, Inc.

Katz, S (1984) *Classic Plastics.* London: Thames & Hudson.

Kornerup, A and Wanscher, J H (1978) *Methuen Handbook of Colour.* London: Methuen.

Kueppers, H (1978) *The Basic Law of Color Theory (Das Grundgesetz Der Farbenlehre).* Cologne: DuMont Buchverlag GmbH & Co.

Lenclos, J and D (1982) *Les Couleurs de la France.* Paris: Editions du Moniteur.

Lindsay, K and Vergo, P (Eds) (1982) *Kandinsky: Complete Writings on Art.* London: Faber & Faber.

Luescher, M (1970) *The Luescher Color Test*. London: Cape.

Munsell, A H (1963) *A Color Notation*. Baltimore: Munsell Color Co.

Naylor, G (1968) *The Bauhaus*. London: Studio Vista.

Ogden, F (1989) Colour Marketing Group International Conference. Toronto.

Russell, D (1990) *Colorworks*. (Volumes 1-5: *The Red Book, The Blue Book, The Yellow Book, The Pastels Book, The Black and White Book.)* Cincinnati: North Light Books.

Sparke, P (1986) *Did Britain Make it?* London: The Design Council.

Stokes, A (1937) *Colour and Form*. London: Faber & Faber.

Varley, H (Ed) (1980) *Colour*. London: Mitchell Beazley.

von Bezold, W (1876) *The Theory of Color in its Relation to Art and Industry*. Boston: L Prang & Co.

von Goethe, J W and Judd, B (1971) *Theory of Colours*. Cambridge: MIT Press.

Whitford, F (1984) *Bauhaus*. London: Thames & Hudson.

Wilson, R F (1960) *Colour in Industry Today: A Practical Book on the Functional Use of Colour*. London: Allen & Unwin.

The Colour Group
The Studio
32-34 Great Marlborough Street
London W1V 1HA
Tel: 071-734 3248   Fax: 071-734 6825

This is a forum created for the exchange of colour information and trends in industrial, commercial and environmental design through workshops and seminars.